Fundamentals of Engineering

A Project-Based and Student-Centered Approach

Tsung-Chow Su

Florida Atlantic University

Contents

Part 1
Engineering Fundamentals

Chapter 1: Introduction of Engineering

A) What is engineering?

Engineering is a profession where the forces of nature are directed to the benefit of humankind. An engineer must understand the characteristics of these forces of nature, and must also understand the needs of the societies they serve; engineers must therefore be educated in science and humanities.

To learn science, one must develop a strong working knowledge of math, which is the language of science. Many people are uncomfortable with math, which prevents them from entering into the engineering profession.

The work of an engineer is directed to benefit humankind; therefore engineers are typically paid well.

Practice Problem 1: Using the internet, determine the top salaries for B.S. degree holders from 15 different fields.

Engineers typically work in teams to address ways to solve various problems. In order for a team to be successful, the members must all possess strong teamwork skills. Communication skills are especially important, as proper communication is the key to keeping the entire team on the same page as they work towards a common goal.

Let's now go back and take a closer look at the difference between a scientist and an engineer. If a scientist finds a law that is not universally valid, it is useless. On the other hand, if an engineer designs something which is universally valid, it is also useless, as engineers must satisfy a specific need according to a specification. Theodore von Karman, a famous 20th century mathematician, physicist, and aerospace engineer once said, "Scientists study the world as it is, engineers create the world that has never been." [1]

Engineering is a human activity aimed at creating new artifacts, algorithms, processes, and systems that serve humanity. Engineers work towards fulfilling the gamut of human needs and wants, such as shelter, food, transportation, communication, security, longevity (personal and for progeny), entertainment, aesthetic pleasure, and social, emotional, spiritual, and psychological rewards.

An engineer should be able to quickly determine how things work, determine what consumers want, create a concept, use math models to improve a concept, build or create prototypes, quantitatively and robustly test prototypes to improve them, determine whether consumer and enterprise values are aligned (engineers should have sound business sense), and communicate all of the

[1] M. David Burghardt, *Introduction to the Engineering Profession, Second Edition*. New York: HarperCollins, 1995. ISBN 0-673-99371-X.

above to various audiences, adjusting the technicality of their communications appropriately based on their audience's knowledge of the topic. Much of this requires "domain-specific knowledge" and experience, several of these skills require systems thinking and statistical thinking, and all require teamwork, leadership, and societal awareness.

Engineering's impact hits you from the moment you wake up. Your alarm clock is run by electricity, and was programmed to go off at a specified time. Your shower faucets are valves which were designed by an engineer to bring the water through your showerhead through pipes connected to a water supply. This entire system was developed by teams of engineers. As you are about to walk out the door, you open the fridge for a quick glass of juice and you throw some bread in the toaster; that refrigerator and toaster were designed according to engineering specifications. It's easy to see how dependent we've become on technology. No longer do we wake with the rising of the sun; the sophistication of using an alarm clock allows us more freedom to schedule our busy lives.

Engineering involves creating and developing products, machines, and systems to benefit humankind. Although some people struggle with creativity, it is something that can be learned and improved upon through practice (Burghardt, *Introduction*, 2).

B) Succeeding in engineering

To succeed as an engineer, one must take math, science, engineering science, and engineering design courses. It is also important to develop a deep knowledge of the world, keeping up with current events and technologies.
A person's achievements in learning something follow an "S" curve based on time and effort. Initially, when someone tries to learn something new, the learning process starts off slowly. As more and more effort is dedicated to learning about it, the quicker they are able to catch on. As effort increases, they become more and more knowledgeable about the subject, and eventually reach a level of diminishing return.

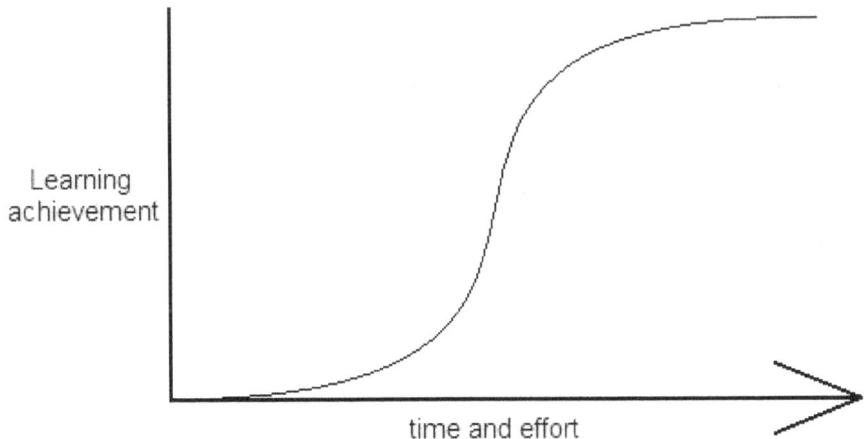

Figure 1: Achievement curve showing that how much you learn about a subject is dependent upon how much time and effort you put into studying that subject.

For example, a "C" paper is merely the first draft of an "A" paper; it is the additional effort that transforms "average" into "excellent".

It is useful to bear in mind that it is impossible to teach a student everything he or she needs to know as preparation for a professional career in a four or five (or even ten) year university program. Engineering is practiced within a much broader societal context - not as an end in itself. Students are taught *how* to learn by developing a fundamental understanding of the unity of fundamental engineering tools and concepts needed for engineering practice (as opposed to providing them with a vast bag of tricks for solving selected problems). These basic fundamentals include mathematics, information technology, science, design and manufacturing, economics and business practices, communication skills (written, oral, graphic, and listening), engineering "design" (creative thinking and open-ended problem solving in the most general sense) and its close connection with manufacturing (i.e., "If you can't build it, you can't use or sell it"), information gathering, critical thinking and evaluation skills, teamwork (not merely "group work") and communication skills, the "Why" and "What" of theory, and how these basics are then applied in practice (the "How" in applications can then be gained by experience and subsequent training).

C) How to plan and study

As the popular saying goes, "no one plans to fail, they only fail to plan." It is important to stay organized and make sure you are prepared for the tasks you are faced with; identify your tasks in and out of school, and set an allotted time to work on said tasks. It is helpful to make use of a weekly chart which outlines the tasks you must complete such as study time, class times, and other important items such as work time obligations.

1) How much time does one need to put into his or her course work?

The rule of thumb is to study three hours a week for each credit hour. For example, if you have a 15-credit course load, you are expected to put in 45 hours a week on school work. For many people, this is challenging. Having a job presents the biggest scheduling challenge, as most workplaces are inelastic with their schedules and you must plan your schooling around that. Students who have jobs must be careful about committing to non-school work, as often study time will get short-changed. When a bad grade comes, it is often too late.

It is important to keep your grades up. You often need a certain minimum GPA in order to get a job, get into graduate school, and get scholarships.

Homework Problem 1: Fill out one of the blank "assignment" pages in this textbook using the example weekly schedule chart below to make notes and plan out your own personal study chart for this semester. Submit your homework online

Time	Monday	Tuesday	Wednesday	Thursday	Friday	Saturday	Sunday
7:00 AM	wake up	wake up	wake up	wake up	wake up	sleep	sleep
8:00 AM	go to FAU	go to FAU	go to FAU	go to FAU	go to FAU	sleep	sleep
9:00 AM	intro to eng	STUDY	intro to eng	STUDY	intro to eng	sleep	sleep
10:00 AM	calc 1	STUDY	calc 1	calc 1	calc 1	wake up	
11:00 AM	english 1	eng graphics	english 1	eng graphics	english 1	eat	wake up
12:00 PM	lunch	eng graphics	lunch	eng graphics	lunch	STUDY	eat
1:00 PM	STUDY	lunch	STUDY	lunch	STUDY	STUDY	STUDY
2:00 PM	STUDY	STUDY	STUDY	STUDY	STUDY	STUDY	STUDY
3:00 PM	STUDY	STUDY	STUDY	STUDY	STUDY	STUDY	STUDY
4:00 PM	break	break	break	break	break	STUDY	STUDY
5:00 PM	STUDY	STUDY	STUDY	STUDY	STUDY	STUDY	STUDY
6:00 PM	STUDY	STUDY	STUDY	STUDY	STUDY	STUDY	STUDY
7:00 PM	free time	free time	free time	free time	free time	free time	free time
8:00 PM	free time	free time	free time	free time	free time	free time	free time
9:00 PM	free time	free time	free time	free time	free time	free time	free time
10:00 PM	free time	free time	free time	free time	free time	free time	free time
11:00 PM	go to bed	go to bed	go to bed	go to bed			

Class Schedule			
Intro to Engineering	MWF	9 - 10	3 credits
Calculus 1	MWRF	10 - 11	4 credits
English Comp 1	MWF	11 - 12	3 credits
Eng. Graphics	T R	11 - 1	3 credits
TOTAL			13 credits
TOTAL STUDY TIME REQUIRED			39 hours

Figure 2: Sample weekly schedule chart.

NAME: _____   DUE: _____

Assignment Title: _____

Course: _____

NAME: _____   DUE: _____

Assignment Title: _____

Course: _____

Chapter 2: Historical Perspective[2]

Civilization as we know it today owes its existence to engineers. These are the people who have learned to exploit the properties of matter and the sources of power for the benefit of humankind. By an organized, rational effort to use the material world around them, engineers devised the comforts and conveniences that mark the difference between our lives and those of our ancestors.

The story of civilization can be thought of as the story of engineering – that arduous struggle to make the forces of nature work for the benefit of humankind. The story of engineering, as well as we can ascertain by piecing together ancient manuscripts and relics, is derived mostly from insight gained from the accounts of kings and philosophers, generals and politicians.

To appreciate the enormous accomplishments of the engineers, it is helpful to gain an understanding of the changes that have taken place in the human way of life over the course of the last million years. A million years ago, our ancestors were small, apelike primates. Two things distinguished our ancestors from modern apes; first, they lived mostly on the ground and tended to walk upright, so their limbs were proportioned similarly to ours. They did not have the long arms, short, bowed legs, and hand-like feet of modern apes; however, their brains were essentially the same as those of modern apes.

Over time, the forces of evolution caused these beings to evolve into early humans, every bit as human in form and as intelligent as we are today. These early humans, like all those who had gone before them, lived by food gathering. They fed themselves by hunting, fishing, picking berries, and digging up edible roots and tubers, and getting protein from lizards, insects, and carrion. Today, only small bands of African Bushmen and Pygmies, a few Australian aborigines, and a handful of Inuit (a tiny fraction of 1% of humanity) continue to survive in this manner.

Because of the difficulty of getting food in 8000 BCE, only a few hundred thousand humans existed on the entire face of the globe. However, there is no reason to think that modern humans are cleverer than these ancestors. For one thing, roughly 10,000 years is too short a time for evolution to have had a measurable effect.

However that may be, humans have spent about 99% of their history, since they first learned to make tools, as a hunting and food-gathering tribesmen. Civilization has arisen only during the remaining 1% of this time, when people discovered how to raise crops and tame animals. These discoveries enabled a square mile of fertile land to support 20 to 200 times as many people as before, and freed some people for specialized occupations.

---

[2] Information in this chapter compiled from L. Sprague De Camp, *The Ancient Engineers: Technology and Invention from the Earliest Times to the Renaissance*. New York: Barnes & Noble Books, 1993, pp 13-27. ISBN 0-88029-456-6.

This revolution seems to have first taken place in the north of Iraq and Syria, where the Agricultural Revolution quickly spread to the Nile and the Indus, which in their turn became centers of culture.

The Agricultural Revolution brought about changes fully as drastic in people's lives as those caused by the more recent Industrial Revolutions. Permanent villages took the place of temporary campsites. One theory suggests that humans were first persuaded to give up their wandering life when they discovered that mashed grass seeds could be used to make beer, since they had to stick around long enough for the mash to ferment.

Over the next three to four thousand years, some of the farming villages of the Middle East grew into cities; as farmers learned to grow more food than they needed to provide for themselves, other people's time was freed up and spent making useful things which they exchanged for the farmer's surplus foods; thus specialization arose.

As specialization increased, merchants, physicians, poets, smiths, and craftsmen of many kinds came into existence. Instead of making their own houses, carts, wells, and boats, people began to buy them from specialists who constructed them.

As chiefs evolved into kings and wizards into high priests, they grew rich and powerful, able to acquire servants who outranked the simple peasants. Slavery, intended at first as a humane means of avoiding having to slaughter one's prisoners of war, introduced still another class. Thus society became seamed and fissured into a multitude of specialized occupations.

Over time, wealth and experience piled up, and people undertook projects too large for a single craftsman, even with the help of his sons and apprentices. These large projects called for the combined efforts of hundreds or even thousands of people. A new class of society arose: technicians or engineers, who could negotiate with a king or priesthood for building a public work, plan the details, and direct the workmen. These people combined practical experience with knowledge of theoretical principles. Sometimes they were inventors as well as contractors, designers, and foremen, but all were people who could imagine something new and transform a mental picture into reality.

Some primitive inventions, like the manioc squeezer of the South American Indians, the Australian Indians, the Australian boomerang and the Inuit toggle-joint harpoon, are extremely ingenious. They point to inventive talents as keen as anything the civilized world can show. Nevertheless, during nearly all of the last million years, invention progressed with extreme slowness. Humans chopped with ax heads held in their fist for hundreds of thousands of years before they learned to fasten handles to their axes.

The reasons for the slowness of invention in primitive societies are not hard to understand. For one thing, primitive peoples live a hand-to-mouth existence. Most of their foods couldn't be stored, so there was never any surplus. Therefore they couldn't really afford to risk experiment like more advanced peoples. If an experiment failed, they died.

As a result, primitive societies were very conservative. Tribal customs prescribe exactly how everything shall be done, on pain of displeasing the gods. An inventor was likely to be considered as a dangerous deviant.

Peasant farmers were almost equally conservative. People's inventive faculties are stimulated by the breakdown of the established custom that takes place in the urban environment; hence most inventions have been made by city dwellers.

Another cause of the slowness of primitive invention is the scarcity of people who could invent. A hunting and food-gathering technology can support only a very small population for a given area. Thus the few hundred thousand members of the human species living at any time before the Agricultural Revolution were divided into many isolated little hunting bands where all members were focused purely on survival, not human progression.

Such bands seldom exceeded fifty to a hundred people, including in this count the many (but short-lived) children. Because the hunters were distance-limited by having to walk to kill their game and carry it back to camp, an increase in numbers did not enlarge the area that can be hunted at one time, but instead merely causes the same area to be hunted more intensively. If a particular band grew too large, game became scarce, and the band was forced to choose between migration and starvation, and so eventually, large bands had to split up.

In any society, only a handful of people have original ideas or make inventions. Of these inventors, only a fraction have the tenacity to continue bettering their inventions until they work, and to promote them until they persuade others to take them up.

A rough idea of the percentage of inventors among modern Americans can be obtained from the statistics of the United States Patent Office, which issues about 40,000 patents every year. We can therefore estimate that the mid-twentieth-century American population of 180,000,000 people produced about one patentable invention each year for every 4,500 citizens. Suppose, now, that all Americans were wiped out except one band of forty-five people like our ancestors' tribes; if this group continued to produce inventions at the same rate, it would turn out only one invention every century! This is an oversimplification, but it indicates why small tribal societies, no matter how clever their tribesmen, could not be expected to produce inventions rapidly. In fact, the rate of inventions among Stone Age hunters was enormously slower than this; modern Americans are encouraged to invent in ways that primitive folk were not. We are used to the thought that people can improve their lot by inventing things, and that invention is a worthy act, whereas primitive people, who exert all their energies on survival, cannot afford to support a fellow tribesman in idleness while he dreams up new ideas and they regard inventors with glowering suspicion.

Suppose that there are two bands of forty-five Americans which are isolated from each other. Each band will produce one invention a century, so that each progresses at the same rate as before. Their cultures will diverge somewhat, as they will only hit upon the same inventions rarely, by chance, but each group will plod along at the same rate of about one invention a century.

However, if the tribes were to meet and join forces, then all of the tribesmen would be able to take advantage of the inventions produced by any one of them. The combined group will produce inventions twice a century instead of once. In other words, they will technologically progress twice as fast. Progress in civilization depends upon invention, and increased rates of invention depend upon sizeable populations that are only possible under civilization.

Once the Agricultural Revolution had taken place, much denser and more numerous populations began to develop around the Nile, the Euphrates, and the Indus. As the Reverend Thomas Malthus pointed out, people quickly breed up to the greatest density the land will support at the current technological level. At that point, the population levels off, because excess people are destroyed by starvation, pestilence, or war.

The inventions on which civilization was founded tended to spread along trade routes to lands where these ideas could be profitably applied. They were stopped by strong, natural barriers such as deserts and oceans, and they died out where conditions made them useless. Raising cotton and dates did not spread to Europe because they would never grow there. The wheel didn't spread from Iraq to Arabia because no primitive wheels were needed in the sands of the desert.

As a result of this spread of technology, a high level of civilization had been achieved 1,000 years before Christ from the lands of the Mediterranean to China. Any new invention that originated at one end only took a few centuries to travel to the other end. Civilization had little effect on northern Europe and northern Asia because the populations of these lands were very thinly scattered.

Geographical barriers prevented civilization from reaching the parts of Africa that lie under the Sahara Desert, the swamps of the White Nile, and the mountains of Abyssinia. This barrier isolated sub-Saharan Africa as effectively as if it had been an island. Civilization also failed to leap the watery barriers to reach the Pacific Islands, Australia, or the Americas at that time.

The main factor in determining whether any people took part in the technological adventure that followed the Agricultural Revolution was simply a matter of geography.

The first engineers were irrigators, architects, and military engineers; the same person was expected to be an expert at all three. Specialization within the engineering profession has developed only over the last few centuries.

Irrigators laid out canal systems on which the early river-valley civilizations depended. Irrigation enabled more farmers to raise more food, which allowed for an increasing number of specialists, relieved of peasant's chores, to gather in the city.

As these cities grew, their kings desired larger houses, more comfortable than the huts of stone, clay, and reeds wherein they had been living. They therefore called upon the architects to build them palaces.

Next, priests insisted that the gods would be offended if they were not housed at least as splendidly as kings. So the architects put up temples containing statues of the gods and other works of art.

To protect the wealth of the gods and kings, military engineers built walls and dug moats around cities.

The hoards of metals, jewels, fine raiment, and food stuff in the temple and palaces also required workers and a means to keep track of it all. Thus came about the invention of arithmetic and writing.

Many ancient writings on stone and clay have survived but the writings on the perishable materials, such as papyrus reeds, palm fronds, and bamboo, have disappeared. As nearly as we can reconstruct the evidence, the early civilizations were patchworks of little independent city-states, ever fighting with one another. In time, the march of technology made the city-state obsolete, taking on various other forms such as limited monarchy, aristocratic republic and popular dictatorship. However, ancient empires continued on strongly and tended to be absolute monarchies of the most despotic kind. The king was God, or at the very least, a special agent of a god. His word was law. Moreover, no one seemed to seriously consider large-scale government of any other kind.

Nowadays we draw fine distinctions among the meanings of such words as craftsperson, engineer, technician, and inventor. In speaking of ancient technical people however, there is no point in observing such delicate differences. We think of an engineer as a person who designs some structure or machine, or who directs the building of it, or who operates and maintains it. In practice, most ancient engineers were inventors while most ancient inventors could also be classed as engineers.

Despite the enormous importance of engineers and inventors in making our daily life what it is, history does not tell much about them. The earliest historical records were made by priests praising their gods and poets flattering their kings. There was not much concern for such mundane matters as technology. Yet interestingly enough, engineering technology has progressed and persevered where governments, religions, the arts, and scientific discoveries have waxed and waned. Barring nuclear war, the end of this fruition of engineering is nowhere in sight.

Practice Problem 2: Using the internet, determine the greatest engineering accomplishments of the 20th century.

Practice Problem 3: Go to
http://www.ted.com/talks/anthony_atala_growing_organs_engineering_tissue.html **to learn about the advancement of engineering human organs.**

Practice Problem 4: Google search and watch the video "Doing more with much less" by Paul MacCready.

Practice Problem 5: Google search and watch the video "Archimedes steam cannon – MIT" and read the article.

Essay Problem 1: Using a blank "assignment" page from this textbook, brainstorm some ideas for an essay explaining why you want to be an engineer. Submit your essay online.

Chapter 3: Fields of Engineering – Disciplines and Careers[3]

The first engineering discipline was military engineering, with their tasks involving serving their rulers by creating stronger armed forces for their realm; all the major early engineering focus was placed on military affairs.

Time passed, and in the 18th century the discipline of civil engineering was born from a need for bridges, canals, and harbors. Further discipline division was born with the coming of the Industrial Revolution; the need for machines and the energy that powered those machines became increasingly important. Mechanical and electrical engineering disciplines were created separately from civil engineering as a result.

When textile and pharmaceutical industrial processes became important, chemical engineering was formed as a new discipline at the turn of the 20th century, and further need for technological advancements led to the birth of computer engineering in the second half of the 20th century.

Thus has the discipline of engineering evolved and subdivided over the years, correspondingly with the advancements in science and technology and the needs of society.

The most common engineering disciplines are listed as follows:

1) Aeronautical and Aerospace Engineering
2) Agricultural Engineering
3) Architectural Engineering
4) Automotive Engineering
5) Biomedical Engineering
6) Ceramic Engineering
7) Chemical Engineering
8) Civil Engineering
9) Computer Science and Computer Engineering
10) Electrical and Electronics Engineering
11) Environmental Engineering
12) Industrial Engineering
13) Manufacturing Engineering
14) Marine Engineering, Naval Architecture, and Ocean Engineering
15) Materials and Metallurgical Engineering
16) Mechanical Engineering
17) Mining and Geological Engineering
18) Nuclear Engineering
19) Petroleum Engineering

Not all disciplines exist in any given university, and some specialized disciplines can only be found at a limited number of universities, such as

---

[3] Information in this chapter compiled from M. David Burghardt, *Introduction to the Engineering Profession, Second Edition*. New York: HarperCollins, pp 46-58, 1995. ISBN 0-673-99371-X.

Molecular Engineering, a less common degree program which is offered at the University of Chicago.

Engineers usually direct a technological team composed of scientists, engineering technologists, technicians, and craftspeople in an effort to design new products or systems.

The purpose of an engineering education is to equip creative minds with the mathematical and analytical skills necessary to conceive new designs, to intelligently question present ways of accomplishing tasks, and to find better alternatives in light of evolving technology.

There are many career paths available to an engineer, including research and development, design, manufacturing, construction, operations, maintenance, quality control, sales, management, government work, consulting, and teaching.

Engineering societies such as the American Society of Civil Engineers (ASCE), the American Society of Mechanical Engineers (ASME), and the Institute of Electrical and Electronics Engineers (IEEE), evolve constantly to serve their discipline's professors, to exchange documented knowledge, to set standards, and to combine efforts with fellow engineers to solve problems within their respective domains.

Practice Problem 6: Research the formation of the ASME, and learn about the birth of the ASME Boiler and Pressure Vessel Code.

Chapter 4: Ethics and Professional Responsibility

With their understanding of nature and humanity, engineers are in a position to influence society, while also being trusted as professionals to produce safe products. Ethics is the application of moral principles, and engineers are expected to be ethical. In the fields of science and math, the paramount ethical consideration is intellectual honesty. Engineering ethics are much broader, and must account for economics, safety, environment, and social impact.

National Society of Professional Engineers Code of Ethics[4]

THE FUNDAMENTAL PRINCIPLES
Engineers uphold and advance the integrity, honor, and dignity of the engineering profession by:
  I. using their knowledge and skill for the enhancement of human welfare;
  II. being honest and impartial, and serving with fidelity the public, their employers and clients;
  III. striving to increase the competence and prestige of the engineering profession; and
  IV. supporting the professional and technical societies of their disciplines.

THE FUNDAMENTAL CANONS
  1. Engineers shall hold paramount the safety, health, and welfare of the public in the performance of their professional duties.
  2. Engineers shall perform services only in the areas of their competence.
  3. Engineers shall issue public statements only in an objective and truthful manner.
  4. Engineers shall act in professional matters for each employer or client as faithful agents or trustees, and shall avoid conflicts of interest.
  5. Engineers shall build their professional reputation on the merit of their services and shall not compete unfairly with others.
  6. Engineers shall act in such a manner as to uphold and enhance the honor, integrity, and dignity of the profession.
  7. Engineers shall continue their professional development throughout their careers and shall provide opportunities for the professional development of those engineers under their supervision.

Please refer to Part 4 of this text for the complete Code of Ethics including suggested ethical guidelines.

The following method, known as the 5 P's, gives a good set of ethical guidelines[5] with the following questions for consideration:

---

[4] Information in this chapter compiled from M. David Burghardt, *Introduction to the Engineering Profession, Second Edition*. New York: HarperCollins, pp 233-234, 1995. ISBN 0-673-99371-X.

[5] Fogler, H. Scott & Steven E. LeBlanc, *Strategies for Creative Problem Solving*. New Jersey: Prentice Hall, 1995, pp 155-156. ISBN 0-13-179318-7.

1. Purpose – what is your objective? Are you comfortable with your objective and what you are trying to accomplish? Can you look at yourself in the mirror knowing you are trying to work towards this purpose?
2. Pride – are you proud of the solution you've developed to complete your objective? Do you feel any doubt or a false sense of pride?
3. Patience – have you taken ample time to consider the consequences of your proposed solution?
4. Persistence – are you sticking to your instincts and making sure you're doing what you feel is right? Can you defend your rationale and ensure your solution is fair and balanced?
5. Perspective – does your solution fit within your personal ideals and belief system? Have you considered how it will fit into the grand scheme of things?

Perspective is the central component – it is the inner guidance you must follow while keeping the other "P's" in mind to ensure you are seeing and thinking clearly and in a way that does not negatively affect your conscience.

Chapter 5: Communication

Since engineers are often required to work in teams, clear, effective communication, whether written or oral, is of paramount importance. The biggest error most people make when communicating is disregarding the receiver of the information. It is your responsibility to ensure that the person receiving information from you is in a position to understand it as you are presenting it. Further, the level of language used is important; the more grandiose and flowery it is, the better chance there is that the information will be misunderstood or disregarded. Be direct and to-the-point; as an engineer you have the ability to even be somewhat blunt in your communications, a certain luxury that cannot be shared, for instance, by philosophy majors.

A) Written Communication[6]

As an engineering student, and even into your engineering profession, you will be expected to deliver written reports. All written reports should contain the following information at the minimum – the abstract, an introduction, a procedural section, a results section, and a discussion/conclusion section. The abstract is a short statement that shares your objective, your results, and your conclusions with the reader. Abstracts are not long and serve to give a quick paraphrase of your report. The introduction is where you state the problem you are addressing and its significance; it is typically only a paragraph or two in length. The procedural section tells the reader what steps you took to get your results. The reader should be able to reproduce your procedure to verify your results if this section is properly completed. The results section summarizes the results you got, often with visual aids such as charts, graphs, and tables. The discussion/conclusion section explains the results and addresses whether the initial problem was solved.

Scientific reports are typically objective and used to present facts in as clear a manner as possible. Avoid subjective language and aim to present yourself logically; technical reports are not like the average persuasive report you will write in your humanities courses. Make sure to address any complications or errors that may have come up when you were running your investigation; for instance, if you were timing something during the experiment and your timer was not very accurate, it is worth mentioning. If your results seem odd to you, it is a good idea to try and explain why this may be.

Another important type of written communication is memo writing (Burghardt, *Introduction*, p 124). Memos are brief and to-the-point, and should include the following four basic elements:
1) Why you are writing the person;
2) What your purpose is;
3) What you want the person to do;
4) When you want it done.

---

[6] Anne Eisenberg, *A Beginner's Guide to Technical Communication*. McGraw-Hill's BEST, 1998, pp 1-19 & 45-58. ISBN 0-07-092045-1.

A third very important piece of written communication is your resume, or vita. This is how you will present yourself and your qualifications to others. A basic sample resume is presented on the following page, created by a female student graduating from the Florida Atlantic University Electrical Engineering department.

JANE DOE

OBJECTIVE

To secure a challenging entry-level position in the field of Electrical Engineering with an interest in RF or DSP Circuitry.

EXPERIENCE

2011–2012   Motorola   Plantation, FL
Intern – Business Light Radio Unit Team

- Worked alongside a team of engineers to assist in the performance of electrical engineering design and analysis of portable product components.
- Co-designed and performed experiments for verification of analysis.
- Presented documented findings.

2010   Boeing   Wichita, KS
Summer Internship – Boeing Defense, Space, and Security (BDS) Program

- Assisted in solving complex engineering problems, performing risk analysis, and providing technical support.
- Prepare reports of test results and recommend design improvements.

2008–2010   ABC Engineering   Pompano Beach, FL
Programmer I

- Created programs using C, C++.
- Read, interpreted, and modified electrical schematics.
- Assisted in the development of technical documents and specifications.

EDUCATION

2008 - 2012   Florida Atlantic University   Boca Raton, FL

- B.S. Electrical Engineering
- Graduated magna cum laude

AFFILIATIONS, INTERESTS, SKILLS

Member of IEEE, computers, RF circuitry, DSP circuitry, ham radio, hardware and software engineering, MATLAB, programming in C, programming in C++

AWARDS

Dean's List every semester

The following pages are an excerpt from "<u>Whitesides Group: Writing a Paper</u>"[7] which covers important topics when it comes to report-writing. As a student, and into your professional career, a well-written report will set you apart and showcase your communication abilities.

WHITESIDES' GROUP: WRITING A PAPER**

By *George M. Whitesides**

1. WHAT IS A SCIENTIFIC PAPER?

A paper is an organized description of hypotheses, data and conclusions, intended to instruct the reader. Papers are a central part of research. If your research does not generate papers, it might just as well not have been done. "Interesting and unpublished" is equivalent to "non-existent".

Realize that your objective in research is to formulate and test hypotheses, to draw conclusions from these tests, and to teach these conclusions to others. Your objective is not to "collect data".

A paper is not just an archival device for storing a completed research program; it is also a structure for *planning* your research in progress. If you clearly understand the purpose and form of a paper, it can be immensely useful to you in *organizing* and conducting your research. A good outline for the paper is also a good plan for the research program. You should write and rewrite these plans/outlines throughout the course of the research. At the beginning, you will have mostly plan; at the end, mostly outline. The continuous effort to understand, analyze, summarize, and reformulate hypotheses on paper will be immensely more efficient for you than a process in which you collect data and only start to organize them when their collection is "complete".

2. OUTLINES

2.1 The reason for outlines

I emphasize the central place of an outline in writing papers, preparing seminars, and planning research. I especially believe that for you, and for me, it is most *efficient* to write papers from outlines. An *outline* is a written plan of the organization of a paper, *including* the data on which it rests. You should, in fact, think of an outline as a carefully organized and presented set of data, with attendant objectives, hypotheses, and conclusions, rather than an outline of text.

An outline itself contains little text. If you and I can agree on the details of the outline (that is, on the data and organization), the supporting text can be assembled fairly easily. If we do *not* agree on the outline, any text is useless. Much of the *time* in writing a paper goes into the text;

---

* The text based on a handout created on October 4, 1989.

most of the *thought* goes into the organization of the data and into the analysis. It can be relatively efficient in time to go through several (even many) cycles of an outline before beginning to write text; writing many versions of the full text of a paper is slow.

All writing that I do – papers, reports, proposals (and, of course, slides for seminars) – I do from outlines. I urge you to learn how to use them as well.

2.2. How should you construct an outline?

The classical approach is to start with a blank piece of paper, and write down, in any order, all important ideas that occur to you concerning the paper. Ask yourself the obvious questions: "Why did I do this work?"; "What does it mean?"; "What hypotheses did I mean to test?"; "What ones did I actually test?"; "What were the results? Did the work yield a new method of compound? What?"; "What measurements did I make?"; "What compounds? How were they characterized?". Sketch possible equations, figures, and schemes. It is essential to try to get the major ideas. If you start the research to test one hypothesis, and decide, when you see what you have, that the data really seem to test some other hypothesis better, don't worry. Write them both down, and pick the best combinations of hypotheses, objectives, and data. Often the objectives of a paper when it is finished are different from those used to justify starting the work. Much of good science is opportunistic and revisionist.

When you have written down what you can, start with another piece of paper and try to organize the jumble of the first one. Sort all of your ideas into three major heaps (1–3).

1. *Introduction*

   Why did I do the work? What were the central motivations and hypotheses?

2. *Results and Discussion*

   What were the results? How were compounds made and characterized? What was measured?

3. *Conclusions*

   What does it all mean? What hypotheses were proved or disproved? What did I learn? Why does it make a difference?

Next, take each of these sections, and organize it on yet finer scale. Concentrate on organizing the *data*. Construct figures, tables, and schemes to present the data as clearly and compactly as possible. This process can be slow – I may sketch a figure five to ten times in different ways trying to decide how it is most clear (and looks best aesthetically).

Finally, put everything – outline of sections, tables, sketches of figures, equations – in good order.

When you are satisfied that you have included *all* the data (or that you know what additional data you intend to collect), and have a plausible organization, give the outline to me. Simply indicate where missing data will go, how you think (hypothesize) they will look, and how you will interpret them if your hypothesis is correct. I will take this outline, add my opinions, suggest changes, and return it to you. It usually takes four to five iterations (often with additional experiments) to agree on an outline. When we *have* agreed, the data are usually in (or close to) final form (that is, the tables, figures, etc., in the outline will be the tables, figures, ... in the paper).

You can then start writing, with some assurance that much of your prose will be used.

The key to efficient use of your and my time is that we start exchanging outlines and proposals as early in a project as possible. *Do not, under any circumstances, wait until the collection of data is "complete" before starting to write an outline.* No project is ever complete, and it saves enormous effort and much time to propose a plausible paper and outline as soon as you see the basic structure of a project. Even if we decide to do significant additional work before seriously organizing a paper, the effort of writing an outline will have helped to guide the research.

2.3. The outline

What an outline should contain:

1. *Title*

2. *Authors*

3. *Abstract*

    Do *not* write an abstract. That can be done when the paper is complete.

4. *Introduction*

    The first paragraph or two should be written out completely. Pay particular attention to the opening sentence. Ideally, it should state concisely the objective of the work, and indicate why this objective is important.

    In general, the Introduction should have these elements:

- The *objectives* of the work.

- The *justification* for these objectives: Why is the work important?

- *Background*: Who else has done what? How? What have *we* done previously?

- *Guidance to the reader*: What should the reader watch for in the paper? What are the interesting high points? What strategy did we use?

- *Summary/conclusion*: What should the reader expect as conclusion? In advanced versions of the outline, you should also include all the sections that will go in the Experimental section (at the level of paragraph subheadings) and indicate what information will go in the Microfilm section.

5. *Results and Discussion*

The results and discussion are usually combined. This section should be organized according to major topics. The separate parts should have subheadings in boldface to make this organization clear, and to help the reader scan through the final text to find the parts of interest. The following list includes examples of phrases that might plausibly serve as section headings:

- Synthesis of Alkane Thiols
- Characterization of Monolayers
- Absolute Configuration of the Vicinal Diol Unit
- Hysteresis Correlates with Roughness of the Surface
- Dependence of the Rate Constant on Temperature
- The Rate of Self-Exchange Decreases with the Polarity of the Solvent

Try to make these section headings as specific and information-rich as possible. For example, the phrase "The Rate of Self-Exchange Decreases with The Polarity of The Solvent" is obviously longer than "Measurement of Rates", but much more useful to the reader. In general, try to cover the major common points:

- Synthesis of starting materials
- Characterization of products
- Methods of characterization
- Methods of measurement
- Results (rate constants, contact angles, whatever)

In the outline, do not write any significant amount of text, but get all the data in their proper place: Any text should simply indicate what will go in that section.

- Section Headings
- Figures (*with* captions)
- Schemes (with captions and footnotes)
- Equations
- Tables (correctly formatted)

Remember to think of a paper as a collection of experimental results, summarized as clearly and economically as possible in figures, tables, equations, and schemes. The text in the paper serves just to explain the data, and is secondary. The more information can be compressed into tables, equations, etc., the shorter and more readable the paper will be.

6. *Conclusions*

In the outline, summarize the conclusions of the paper as a list of short phrases or sentences. Do not repeat what is in the Results section, unless special emphasis is needed. The Conclusions section should be just that, and not a summary. It should add a new, higher level of analysis, and should indicate explicitly the significance of the work.

7. *Experimental*

Include, in the correct order to correspond to the order in the Results section, all of the paragraph subheadings of the Experimental section.

2.4. In Summary

- Start writing possible outlines for papers *early* in a project. Do not wait until the "end". The end may never come.

- Organize the outline and the paper around easily assimilated data – tables, equations, figures, schemes – rather than around text.

- Organize in order of importance, not in chronological order. An important detail in writing papers concerns the weight to be given to topics. Neophytes often organize a paper in terms of chronology: that is, they give a recitation of their experimental program, starting with their cherished initial failures and leading up to a climactic successful finale. *This approach is completely wrong. Start with the most important results*, and put the secondary results later, if at all. The reader usually does not care how you arrived at your big results, only what they are. Shorter papers are easier to read than longer ones.

3. SOME POINTS OF STYLE

- Do not use nouns as adjectives:
  Not:
  ATP formation; reaction product
  But:
  formation of ATP; product of the reaction

- The word "this" must always be followed by a noun, so that its reference is explicit.
  Not:
  This is a fast reaction; This leads us to conclude
  But:
  This reaction is fast; This observation leads us to conclude

- Describe experimental results uniformly in the past tense.
  Not:
  Addition of water *gives* product.
  But:
  Addition of water *gave* product.

- Use the active voice whenever possible.
  Not:
  It was observed that the solution turned red.
  But:
  The solution turned red. *or*
  We observed that the solution turned red.

- Complete all comparisons.
  Not:
  The yield was higher using bromine.
  But:
  The yield was higher using bromine than chlorine.

- Type all papers double-spaced (not single- or one-and-a- half-spaced), and leave two spaces after colons, and after periods at the end of sentences. Leave generous margins

- Assume that we will write all papers using the style of the American Chemical Society. You can get a good idea of this style from three sources:

- *The journals.* Simply look at articles in the journals and copy the organization you see there.

- *Previous papers from the group.* By looking at previous papers, you can see exactly how a paper should "look". If what you wrote looks different, it probably is not what we want.

- *The ACS Handbook for Authors.* Useful, detailed, especially the section on references, pp. 173–229.

I also suggest you read Strunk and White, *The Elements of Style* (Macmillan: New York, 1979, 3rd ed.) to get a sense for usage. A number of other books on scientific writing are in the group library; these books all contain useful advice, but are not lively reading. There are also several excellent books on the design of graphs and figures.

B) Oral Communication

When it comes to oral communication, it is important to know your audience and adjust the technical content of your presentation accordingly in order to attract the audience and keep their attention.

Refer to the link below to help develop your oral communication skills by learning how to create an "elevator speech" – the idea is to communicate everything pertinent about yourself in the time it takes an elevator to travel from the bottom floor to the top (about 200 words), and sound good doing so. [8]

http://www.brassmagazine.com/article/elevator-talk-30-second-bio-get-you-remembered

---

[8] Amy Drew, "Elevator Talk – a 30-second bio to get you remembered," written October 30, 2006.

Chapter 6: Basic Mathematics

Engineers deal with numbers and quantitative information. Mathematics plays an important role in engineering analysis. Engineers should have a good working knowledge of mathematics; a comprehensive set of basic engineering mathematics review exercises is included in this section.

A) Two-point Formula

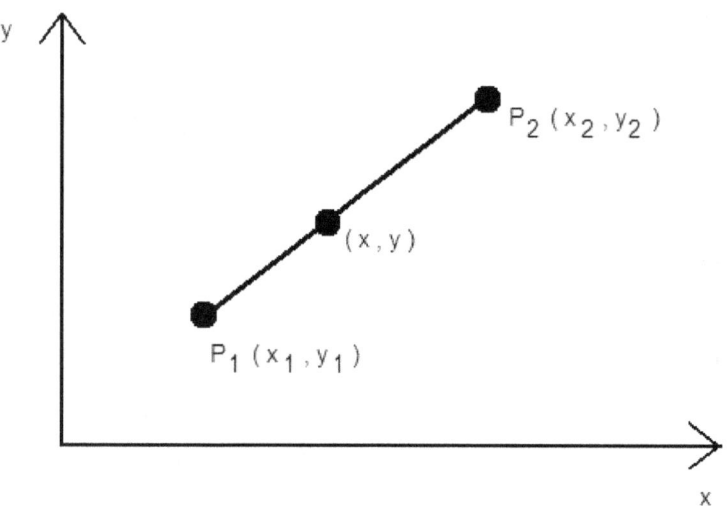

$$\frac{y - y_1}{x - x_1} = \frac{y_2 - y_1}{x_2 - x_1}$$

$$y = y_1 + \left(\frac{y_2 - y_1}{x_2 - x_1}\right)(x - x_1)$$

B) Distributive Property

The distributive property states that
$a (b + c) = a b + a c$

For example, $3 (4 + 2) = 3 * 4 + 3 * 2$
(remember your order of operations!)

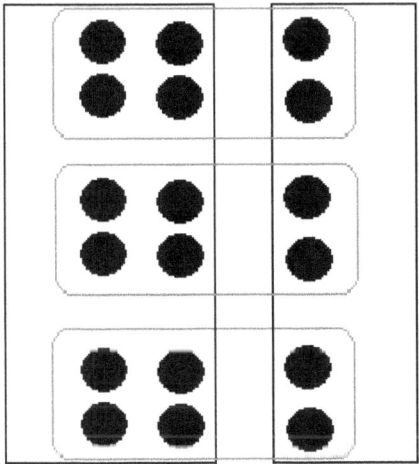

Another example is $6 (5 x + 2) = 30 x + 12$

Note that division is the inverse of multiplication.
Therefore, $\frac{1}{a} (b + c) = \frac{b}{a} + \frac{c}{a}$

C) Math Test

<u>Polynomials and Factoring</u>

1. $x^3 \cdot x^5 =$

2. $(2x^2 + 3x + 8) - (5 - 2x + 2x^2) =$

3. $(x^2 + 3)(2x^2 - x + 2) =$

4. $(3 + 5x)(3 - 5x) =$

5. $(3 + 5x)^2 =$

6. Factor: $5bc + 25b$

7. Factor: $x^2 - 5x - 14$

8. Factor: $27x^3 - y^3$

9. Factor: $xy + 2y^2 + 2xz + 4yz$

10. Factor completely: $x^4 - 16$

<u>Exponents, Radicals, and Complex Numbers</u>

1. $(4a^2b^3)^2(2a^3b)(b^2)^{-2}(c^0) =$

2. $\dfrac{x^{2n+1}}{x^n} =$

3. (a) 2^3 (b) 2^{-3} (c) $4^{3/2}$ (d) $4^{-3/2}$ (e) $144^{1/2}$

4. (f) $(-25)^{1/2}$ (g) $(-1)^{1/3}$

5. Does $\sqrt{a} + \sqrt{b} = \sqrt{a + b}$? Why or why not?

6. Does $a^2 + b^2 = (a + b)^2$? Why or why not?

7. Rationalize the denominator: $\frac{-9xy^3}{\sqrt{3xy}}$

8. Rationalize the denominator: $\frac{4}{\sqrt{x}-\sqrt{y}}$

9. Factor and simplify: $(x^2 + 1)^{1/3} - (x^2 + 1)^{4/3}$

10. $(3 + 2i)(2 - i) =$

11. Write $\frac{3+2i}{2-i}$ as a complex number.

12. Solve $x + 3i = 6 - yi$ for x and y, where x and y are real numbers.

13. What is $i^{1/2}$? Hint: $e^{i\theta}$

Equations and Inequalities

1. Solve for a: $ax + by = c$ Assume $x \neq 0$.
   Why is it convenient to assume this?

2. Solve for a: $ax + by > c$, when (1) $x > 0$ (2) $x < 0$ (3) $x = 0$.

3. Solve: $3 + 2x = 4 + 2x$

4. Solve: $-3 \leq 1 - 2x < 6$

5. Solve: (a) $|2x - 7| = 3$ (b) $|2x - 7| < 3$ (c) $|2x - 7| \geq 3$.

6. Solve $x^2 \leq -3x + 4$ and graph the solution on a number line.

D) Fractions

<u>Simplify the following fractions:</u>

1) $\dfrac{a}{b} + \dfrac{c}{b}$

2) $\dfrac{a}{b} - \dfrac{c}{b}$

3) $\dfrac{a}{b} + \dfrac{c}{d}$

4) $\dfrac{a}{b} - \dfrac{c}{d}$

5) $\left(\dfrac{a}{b}\right)\left(\dfrac{c}{d}\right)$

6) $\dfrac{\dfrac{a}{b}}{\dfrac{c}{d}}$

7) $\dfrac{\dfrac{a}{b} + \dfrac{c}{d}}{\dfrac{e}{f} + \dfrac{g}{h}}$

8) $\dfrac{ac}{bc^2}$

Add and subtract the following fractions (common denominator):

9) $\dfrac{1}{3} + \dfrac{2}{3}$

10) $\dfrac{1}{4} + \dfrac{2}{4}$

11) $\dfrac{1}{5} + \dfrac{8}{5}$

12) $1\dfrac{1}{4} + \dfrac{2}{4}$

13) $\dfrac{2}{3} - \dfrac{1}{3}$

14) $\dfrac{10}{5} - \dfrac{8}{5}$

15) $\dfrac{9}{12} - \dfrac{7}{12}$

16) $3\dfrac{1}{12} - 1\dfrac{11}{12}$

Add and subtract the following fractions (no common denominator):

17) $\dfrac{3}{5} + \dfrac{1}{6}$

18) $\dfrac{1}{2} - \dfrac{1}{3}$

19) $\dfrac{3}{4} + \dfrac{7}{6}$

20) $3\dfrac{1}{4} - 1\dfrac{2}{3}$

21) $3 - 2\dfrac{3}{5}$

Reduce the following fractions:

22) $\dfrac{2}{4}$

23) $\dfrac{3}{3}$

24) $1\dfrac{3}{6}$

25) $\dfrac{10}{8}$

26) $\dfrac{10}{15}$

Multiply and divide the following fractions:

27) $\dfrac{1}{3} * \dfrac{2}{3}$

28) $\dfrac{9}{7} * \dfrac{3}{10}$

29) $\dfrac{1}{3} \div \dfrac{3}{10}$

30) $\dfrac{9}{7} \div \dfrac{2}{3}$

31) $\dfrac{2}{3} \div \dfrac{9}{7}$

Multiply and divide the following fractions using cancellation:

32) $\dfrac{1}{3} * \dfrac{3}{2}$

33) $\dfrac{121}{84} * \dfrac{77}{132}$

34) $\dfrac{72}{45} * \dfrac{54}{8}$

35) $\dfrac{1000}{890} \div \dfrac{500}{89}$

36) $\dfrac{81}{2048} \div \dfrac{27}{1024}$

37) $\dfrac{10}{11} \div 5$

38) $2 \div \dfrac{4}{5}$

Simplify the following complex expressions:

39) $\dfrac{\dfrac{1}{5} + \dfrac{4}{5}}{\dfrac{3}{20}}$

40) $\dfrac{\dfrac{3}{4} - \dfrac{1}{6}}{\dfrac{1}{3} + \dfrac{2}{5}}$

41) $\dfrac{1}{2} * \dfrac{2}{3} * \dfrac{3}{4} \div 4$

Convert the following improper fractions to mixed numbers:

42) $\dfrac{5}{4}$

43) $\dfrac{6}{4}$

44) $\dfrac{27}{8}$

Convert the following mixed numbers to improper fractions:

45) $7 \dfrac{3}{4}$

46) $2 \dfrac{9}{10}$

47) $1 \dfrac{2}{3}$

E) Logarithms and Trigonometry

Logarithms

1. What is the domain of ln(x)? What is the range?

2. What is the domain of 2^x? What is the range?

3. Solve with logarithms: $3^{10} = 3^{5x}$

4. Solve: $2^7 = (x - 1)^7$

5. Write the equation $\log_h x = y$ in exponential form.

6. Write the equation $b^u = v$ in logarithmic form.

7. $\log_{1/4} 64 =$ _____

8. $\ln(e^2) =$ _____

9. $\log 4 + \log 25 =$ _____

10. $\log_4 8 =$ _____

11. $\log_5 125 =$ _____

12. Solve: $\log_3 x = -2$

13. Solve: $\log_x 81 = 4$

14. $a^{\log_a x} =$ _____

15. $\log_a a^x =$ _____

16. $\log_3 1 =$ _____

17. $\log_3(-1) =$ _____

18. If $\log(a) = 5$, what is $\log(a^{-5})$?

19. $\log 200 - \log 2 =$ _____

20. Solve: $\log_5 (x + 1) = \log_5 25$

21. Solve: $\log_{x-1} 31 = \log_5 31$

22. (calculator, change of base) $\log_2 9 =$ _____

23. (calculator, change of base) $\log_{23} 128 =$ _____

24. Write $\log_a ((x-1)^{-2} * (y+2)^3) / \sqrt{x}$ in terms of simpler logarithmic forms.

25. If $\log_a 1.5 = r$, $\log_a 2 = s$, and $\log_a 5 = t$, find $\log_a [(1.5)^3 \sqrt[5]{\frac{2}{5}}]$

26. Does $\ln x / \ln y = \ln(x - y)$? Why or why not?

27. Solve: $\log_3 (x - 2) = 2 - \log_3 x$.

Trigonometry - Graphing

28. Graph $y = \sin(x)$.

29. Graph $y = \sin(x - \pi/2)$.

30. Graph $y = \cos(x) + 1$.

31. Graph $y = \tan(x)$.

32. Graph $y = \tan^{-1}(x)$.

33. Graph $y = e^{x-1}$.

34. Graph $y = 1/(x+2)$.

35. Graph $y = 2\sin(x) - 1$.

36. Graph $y = \tan(2x)$.

37. Graph y = tan(2x + 2). (Careful!)

Trigonometry – Radians and Degrees

38. What's 90 degrees in radians? 120 degrees? 180 degrees?

39. What's π/6 radians in degrees? π/3? π/4?

40. What's x radians in degrees? x degrees in radians?

Trigonometry – Unit Circle

41. $\sin(\pi/2) =$

42. $\sin(\pi) =$

43. $\cos(3\pi/2) =$

44. $\sin(\pi/6) =$

45. $\tan(\pi/6) =$

46. $\tan(\pi/2) =$

47. $\tan(\pi/4) =$

48. $\cos(3\pi/4) =$

49. $\sin(5\pi/6) =$

50. $\tan(5\pi/4) =$

51. $\cos(5\pi/3) =$

52. $\sin(-\pi/4) =$

53. $\sin^{-1}(1/2) =$

54. $\sin^{-1}(-1/2) =$

55. $\cos^{-1}(-1/2) =$

56. $\sec(\pi/3) =$

57. $\csc(\pi/3) =$

58. $\cot(\pi/3) =$

59. Let θ represent the angle of a point on the unit circle. If $\sin(\theta) > 0$ and $\tan(\theta) > 0$, what quadrant is the point in?

60. What about if $\sin(\theta) < 0$ and $\cos(\theta) > 0$?

61. Are the sine, cosine, and tangent functions odd, even, or neither?

62. If $\sin(\theta) = 12/13$ and θ is in quadrant II, what are $\cos(\theta)$ and $\tan(\theta)$?

Inverse Functions

63. What is the range of the inverse sine function? Cosine? Tangent?

64. Find $\sin^{-1}(\cos \pi)$.

65. Find $\cos(\arctan(4/3))$.

Trig identities (memorize these or be able to derive them)

66. $\sin^2 x + \cos^2 x =$

67. $1 + \tan^2 x =$

68. $1 + \cot^2 x =$

69. $\sin(x + y) =$

70. $\cos(x + y) =$

71. $\sin(2x) =$

72. $\cos(2x) =$

73. $\sin(x/2) =$

74. $\cos(x/2) =$

75. Show that $1 + \tan^2 x = 1 / \cos^2 x$

76. Show that $\sin(x)\sec(x) = \tan(x)$.

77. Simplify $\sin^2(x) + \sin^2(x)\tan^2(x)$.

78. Show that $1/(1-\sin(x)) + 1/(1+\sin(x)) = 2 \sec^2(x)$.

79. Show that cos(x)/(1-sin(x)) = sec(x) + tan(x).

80. Find cos15 degrees (no calculator, use cos(x − y))

81. Find cos(5π/12).

82. Show that sin(x + 2π) = sin(x).

83. Express cos3t in terms of sint and cost.

84. Find all solutions of the equation cost = 0.

85. Find all solutions of the equation $\tan\theta\cos^2\theta - \tan\theta = 0$.

86. Find all solutions of the equation sin2θ − 3sinθ = 0 in the interval [0, 2π).

87. Find all solutions of the equation cos3x = 0.

Law of cosines, law of sines - calculators allowed

88. State the law of cosines.

89. State the law of sines.

90. In the following problems let a, b, c be the side lengths of a triangle and α, β, γ be the angles opposite the sides. Find β if a = 10, b = 15, c = 21.

91. Find c if a = 10, b = 12, γ = 108 degrees.

92. Prove that if ABC is a right triangle, the law of cosines reduces to the Pythagorean theorem.

93. Find the other parts of the triangle if α = 38 degrees, β = 64 degrees, and c = 24.

94. (tricky) Find β if α = 60 degrees and a = 5.

Functions, etc.

95. Let f be an odd function. If f (3) = 5, then what is f (−3)? What is −f (3)? What is −f (−3)?

96. Let f be an even function. If f (3) = 10, then what is f (−3)?

97. A central angle θ subtends an arc of length 12 inches on a circle whose radius is 6 inches. Find the radian measure of the central angle.

98. A math professor walks towards the university clock tower and wants to find the height of the clock. She determines the angle of elevation to be 30 degrees, and after proceeding an additional 60 feet towards the base of the tower, finds the angle of elevation to be 40 degrees. What is the height of the clock tower?

F) Reading a graph

Reading data from a graph is an important basic engineering skill. Practice finding the friction factor from the Moody chart below[9]; Moody charts play a central role in pipe design.

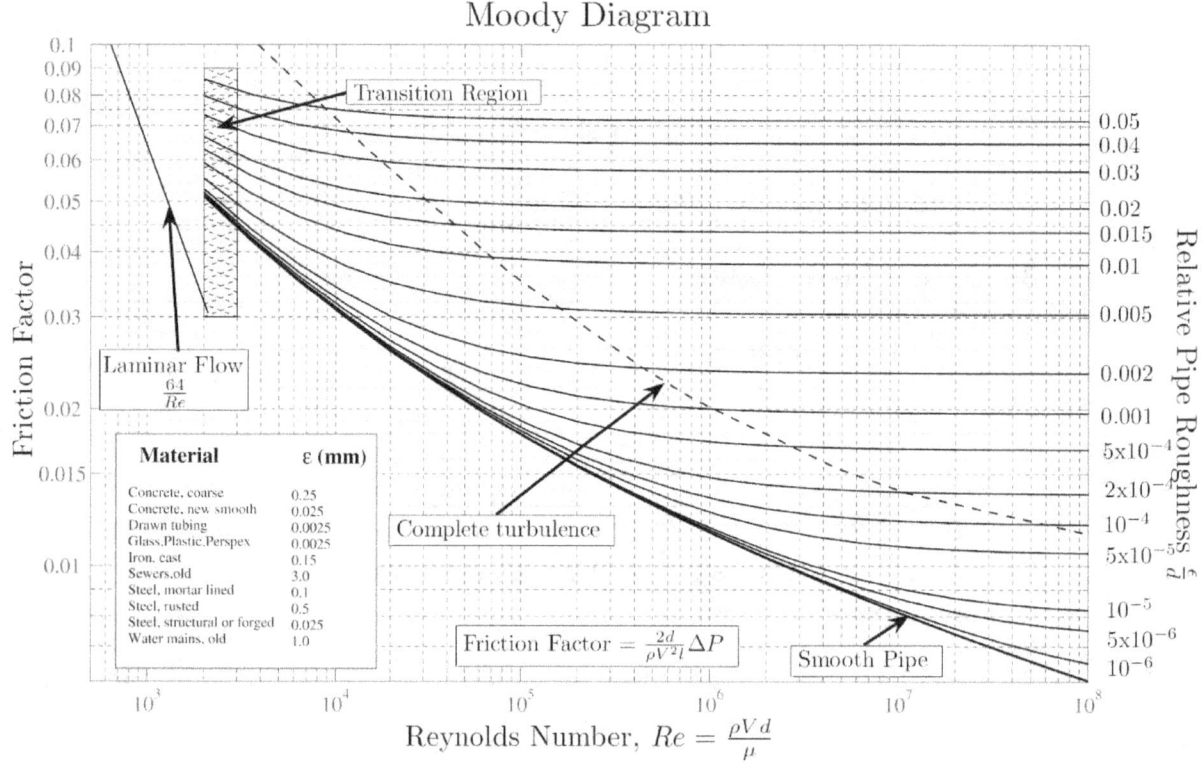

Figure 4: Moody diagram

---

[9] Moody, L. F., *Friction factors for pipe flow*, Transactions of the ASME 66 (8): pp 671 – 684, 1944.

Chapter 7: Using Mathematics to Solve Problems

A boat is adrift in the ocean; the wind speed is U_1 and the ocean current speed is U_2 as shown in the figure below. What is the boat's speed U_B?

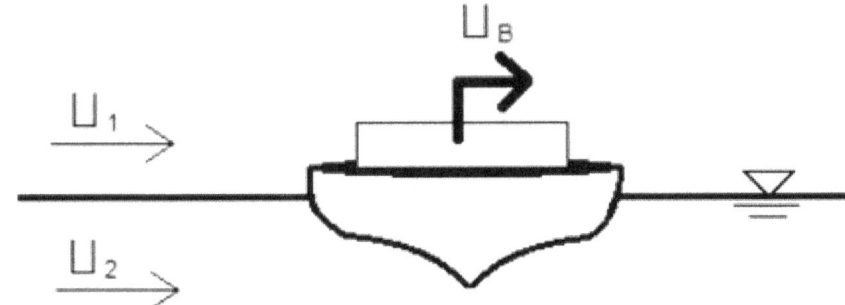

The force on the boat from the wind: $\quad F_1 = C_1 (U_1 - U_B)^2$
The force on the boat from the current: $\quad F_2 = C_2 (U_B - U_2)^2$

In a steady drift state, the two forces acting on the boat will be equal:
$$C_1 (U_1 - U_B)^2 = C_2 (U_B - U_2)^2$$

We can say that: $f = (C_1 / C_2)^{\frac{1}{2}}$

Thus, the boat's speed is:

$$U_B = \frac{1}{(1+f)} U_2 + \frac{f}{(1+f)} U_1$$

Practice Problem 7: Use Google to research "SS Central America Mathematical Treasure Hunting" to see how this formula was used to locate a lost ship.

Chapter 8: Engineering Design

Design is a passionate pursuit! One must first identify a particular need, and then come up with a specification to satisfy the need. With a proper set of specifications, one may proceed to brainstorm to come up with ideas that could meet the specification.

An idea in its infancy is a fragile thing. One must encourage its growth, adding to it, modifying it, and transforming it instead of criticizing and diminishing it.

There are five "F's" of engineering design:
1) Form: What will the design look like? Free-hand sketches are helpful.
2) Function: How will it work?
3) "Fysics" (physics): What is the science behind the design?
4) Flow: The energy flow and mass flow must match in order for the design to function properly.
5) Fabrication: How will it be made?

After one has completed the task of identifying the 5 "F's" for a particular design, the benefit/cost issue and impact must be addressed.

Gathering information is of crucial importance when developing a new design. Practice researching patent information by visiting www.uspto.gov. [10]

When you are proposing a design, you must be confident and enthusiastic. If you are not enthusiastic about your own design, others will be less likely to support it.

---

[10] "United States Patent and Trademark Office," Last modified May 7, 2012, http://www.uspto.gov/

Part 2: Team Projects, Midterm and Final

Project 1: Newspaper Bridge

Description of project – Students will team up to design a bridge with a 5 foot span using only newspaper and one roll of tape. The bridge must be able to support hardcover textbooks which will be placed in the center of the bridge. The team's grade depends on the number of hard-cover textbooks that the bridge is able to withstand at mid-span. This project will help students learn about basic statics, the mechanics of beam-bending, and how to make a fragile material strong (as form determines the strength of a material). Each group will bring their design to class for a competition where they learn from one another what design works best, which designs do not work as well, and why.

This project requires the submittal of a 500-word (minimum) report.

Project 1: Newspaper Bridge

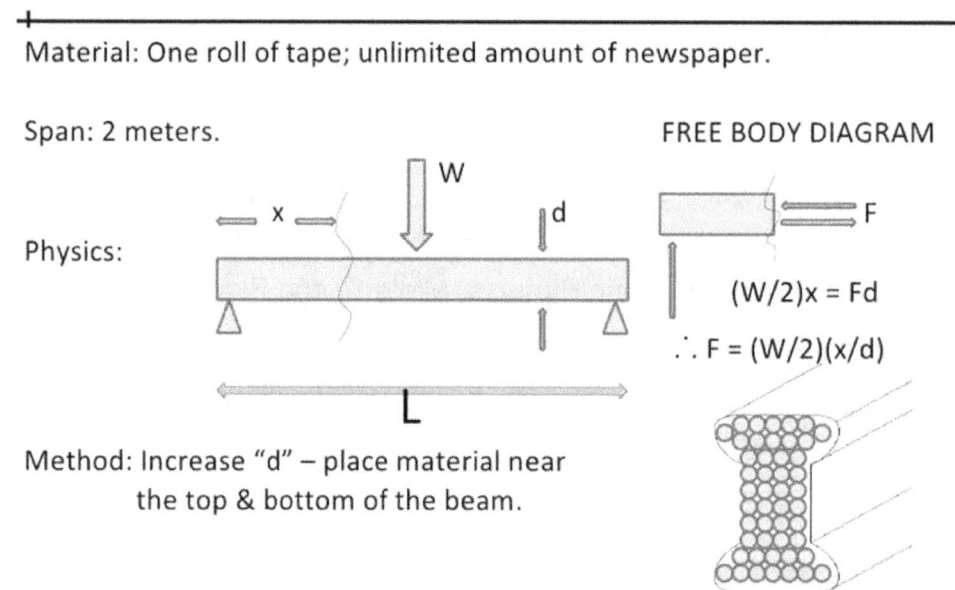

Material: One roll of tape; unlimited amount of newspaper.

Span: 2 meters.

Physics:

FREE BODY DIAGRAM

$(W/2)x = Fd$

$\therefore F = (W/2)(x/d)$

Method: Increase "d" – place material near the top & bottom of the beam.

Grading Scale

Bridge supports 1 book 80%   Bridge supports 2 books 85%   Bridge supports 3 books 90%

Bridge supports 4 books 92%   Bridge supports 5 books 94%   Bridge supports 6 books 96%

Bridge supports 7 books 98%   Bridge supports 8 books 100%

Project 2: Egg Drop Contest

Description of project – Students will team up to drop an egg from the top of a building down to a concrete floor with only newspaper to protect the egg. Students may only use newspaper, no tapes or adhesives. The grade depends on how many stories high the egg is dropped from and how many pages of newspaper are used in the design. This project will help students learn about the physics behind forces upon impact and how to minimize these forces.
This project requires the submittal of a 500-word (minimum) report.

Project 2: Egg Drop Contest

Material: 2 pieces of newspaper or less; grade depends on amount used and height of drop.

Physics: $F = ma = m \left(dV/dt \right)$

Method: Reduce velocity, increase duration of impact, or do both.

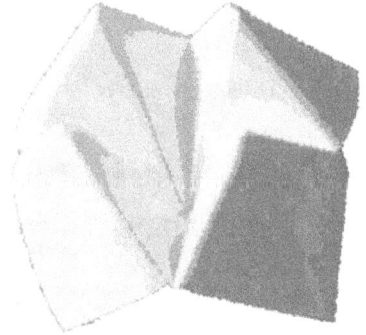

Grading Scale

Drop from the 2nd floor; use 2 full pages of newspaper	80%
Drop from the 2nd floor; use 1 ½ full pages of newspaper	85%
Drop from the 2nd floor; use 1 full page of newspaper	90%
Drop from the 2nd floor; use ½ of a full page of newspaper	100%
Drop from the 2nd floor; use ¼ of a full page of newspaper	110%

10 bonus points will be given to teams that safely drop their egg from the 3rd floor.

Project 3: Paper Glider

Description of project – Students team up to make a paper glider using an index card and a paper clip. The grade will be proportional to the horizontal distance the index card glides. Students will learn about lift/drag, directional control, and experimental uncertainty using a very simple device.
This project requires the submittal of a 500-word (minimum) report.

Project 3: Paper Glider

Materials: with an index card and paperclip, the horizontal distance/ vertical drop will determine the grade.

Physics:

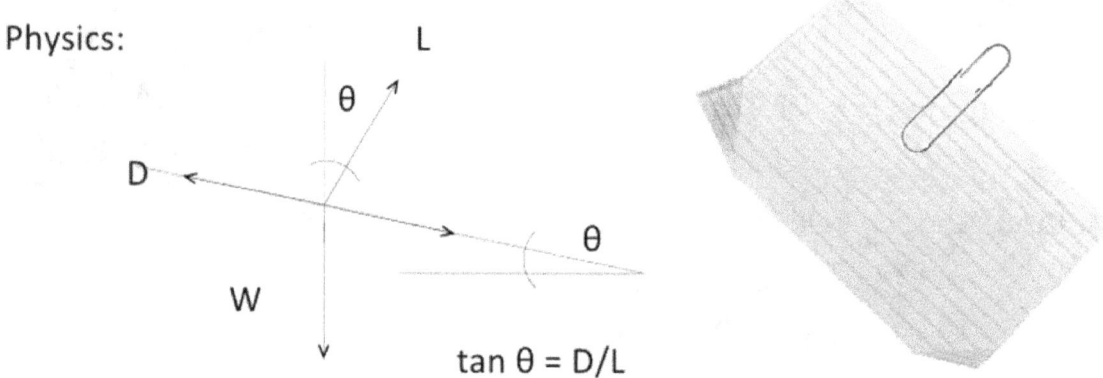

$$\tan \theta = D/L$$

Method: Depending on the position of the paper clip, center of gravity shift and flight path varies. Path is a function of initial angle. This project shows the importance of practice, even for a simple task.

Grading Scale

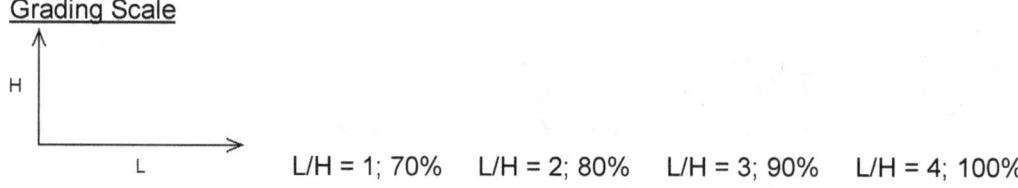

L/H = 1; 70%   L/H = 2; 80%   L/H = 3; 90%   L/H = 4; 100%

Project 4: Solar-Powered Cooker

Description of project – Students team up to design a solar cooker that fits in a 1x3m space to cook water in a baby bottle for 20 minutes. The grade received will be proportional to the temperature achieved. Students will learn about basic physics of heat transfer and use careful reasoning to achieve working designs, and about the uncertainty involved in the performance as this project depends on weather conditions.

This project requires the submittal of a 500-word (minimum) report.

Project 4: Solar-Powered Cooker

Materials: in a 1x1x1 meter cube size limit, cook a small jar of water. Highest temperature achieved in 20 minutes determines the highest grade.

Physics:

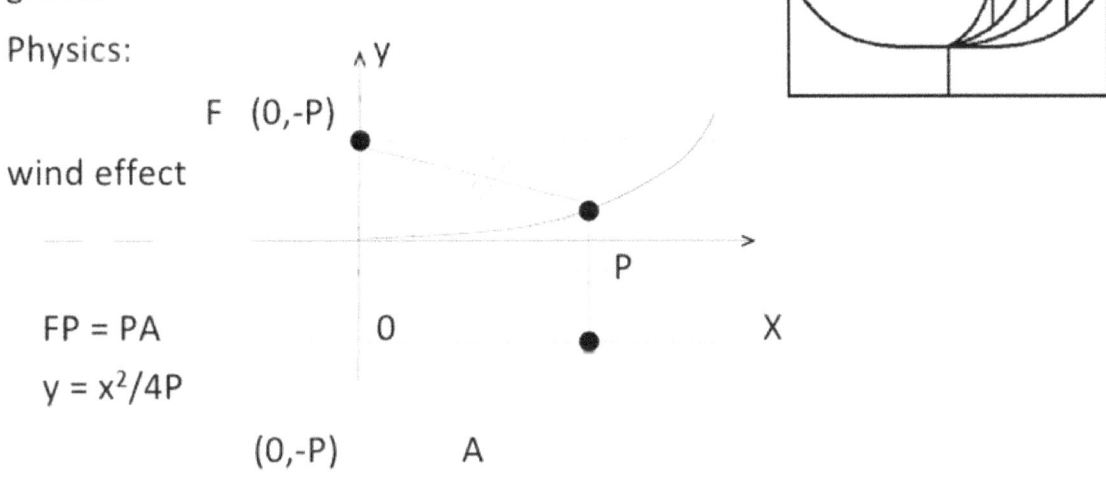

F (0,-P)

wind effect

FP = PA

$y = x^2/4P$

(0,-P)   A

Grading Scale

$70 + (T_{achieved} - T_{ambient})$

Project 5: Solar-Powered Water Distillation System

Description of project – Students team up to come up with a solar-powered water distillation system. The grade will be determined by the amount of water distilled in a 1 cubic meter large distillation system in 40 minutes. Students will learn about basic physics of heat transfer and use careful reasoning to achieve working designs, and also learn about the uncertainty involved in the performance as this project depends on weather conditions.
<u>This project requires the submittal of a 500-word (minimum) report.</u>

Project 5: Solar-Powered Water Distillation System

Materials: in a 1x1x1 meter cube size limit, place a small jar to collect distilled water. Greatest amount collected determines the highest grade.

<u>Grading Scale</u>

Any water collected	70%
2 mL of water collected	80%
6 mL of water collected	90%
10+ mL of water collected	100%

Project 6: Design a Boomerang

Description of project – Students team up to design a boomerang. Students are instructed to throw the boomerang inside of a classroom, have it reach a prescribed distance, and have it return to the student. Students are taught how the shape determines lift, which will provide the centripetal force that allows the boomerang to fly in a circular path. The grade depends on how closely it returns to the thrower. To be successful, they need to learn not only the science, but to integrate it into the way of human operations to assume the fly path is within the confines of the classroom and meets the prescribed conditions.

This project requires the submittal of a 500-word (minimum) report.

Project 6: Design a Boomerang

Materials: Make a boomerang less than 25 cm long with paper and tape. The grade is based on the boomerang's ability to be thrown from the podium of the classroom and have it reach the back of the room then return to the podium. The closest to returning to the podium gets the highest grade.

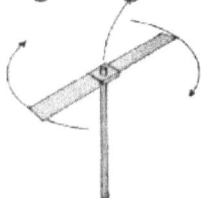
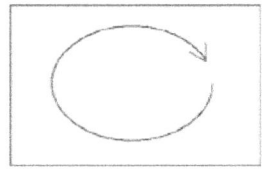
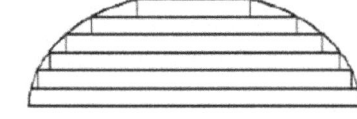

flight path   cross-section of foil

Grading Scale

The highest grade in the class (100%) will be given to the group whose boomerang gets the closest to returning to the podium from which it was launched. The rest of the grades in the class will depend on how close the other boomerangs get to meeting that distance.

Project 7: Design a Catapult

Description of project – Students will design a catapult using a rubber band to project a table tennis ball forward in an enclosed, indoor environment.
<u>This project requires the submittal of a 500-word (minimum) report.</u>

Project 7: Design a Catapult

Materials: Using a rubber band, students will project a table tennis ball in a confined indoor environment. The grade will be determined by how far the table tennis ball is catapulted.

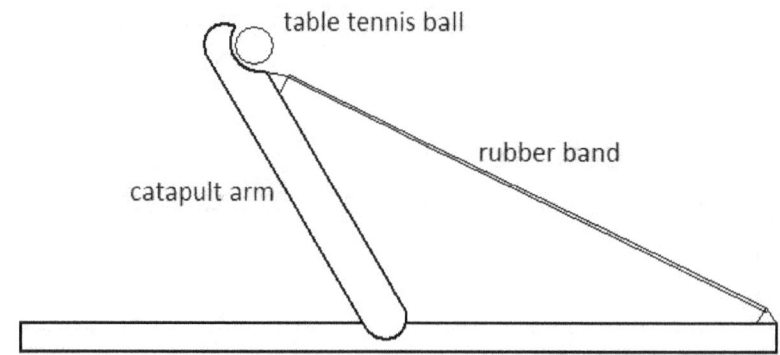

Grading Scale
10'	80%
20'	90%
30'	100%

Project 8: Mousetrap Car Race

Description of project – The eighth and final project is a mousetrap-powered car race. Each group will be issued a mousetrap and use its spring power as an engine in order to build a car for a distance race. The grade will depend on the distance traveled.

This project requires the submittal of a 500-word (minimum) report.

Project 8: Mousetrap Car Race

Use a simple mousetrap to power a car designed by the student; the distance it travels will determine the grade.

Grading Scale

20'	80%
40'	90%
60'	100%

Midterm and Final Examinations

Design Proposal Presentations I and II will serve as midterm and final examinations.

Students will be asked to find a topic of particular interest to them and design a solution. Students must identify the need for their design, the market, the design specifications, and show its Form, Function F(ph)ysics, Flow, and Fabrication (five "F's").

Students will deliver an oral presentation and write a report for each of these proposal presentations.

Peer Evaluations

Near the end of the semester, there will be a peer evaluation session.

Students will be asked to practice their professionalism by distributing appropriate credit points within their team and justify their decision with comments on the evaluation form they are given.

Evaluation results are confidential, but we believe students need to be able to make unbiased, professional, ethical judgment calls regarding their peer's performance.

Part 3: Code of Ethics

Preamble

Engineering is an important and learned profession. As members of this profession, engineers are expected to exhibit the highest standards of honesty and integrity. Engineering has a direct and vital impact on the quality of life for all people. Accordingly, the services provided by engineers require honesty, impartiality, fairness, and equity, and must be dedicated to the protection of the public health, safety, and welfare. Engineers must perform under a standard of professional behavior that requires adherence to the highest principles of ethical conduct.

I. Fundamental Canons

Engineers, in the fulfillment of their professional duties, shall:

1. Hold paramount the safety, health, and welfare of the public.
2. Perform services only in areas of their competence.
3. Issue public statements only in an objective and truthful manner.
4. Act for each employer or client as faithful agents or trustees.
5. Avoid deceptive acts.
6. Conduct themselves honorably, responsibly, ethically, and lawfully so as to enhance the honor, reputation, and usefulness of the profession.

II. Rules of Practice

1. Engineers shall hold paramount the safety, health, and welfare of the public.
    a. If engineers' judgment is overruled under circumstances that endanger life or property, they shall notify their employer or client and such other authority as may be appropriate.
    b. Engineers shall approve only those engineering documents that are in conformity with applicable standards.
    c. Engineers shall not reveal facts, data, or information without the prior consent of the client or employer except as authorized or required by law or this Code.
    d. Engineers shall not permit the use of their name or associate in business ventures with any person or firm that they believe is engaged in fraudulent or dishonest enterprise.
    e. Engineers shall not aid or abet the unlawful practice of engineering by a person or firm.
    f. Engineers having knowledge of any alleged violation of this Code shall report thereon to appropriate professional bodies and, when relevant, also to public authorities, and cooperate with the proper authorities in furnishing such information or assistance as may be required.

2. Engineers shall perform services only in the areas of their competence.
    a. Engineers shall undertake assignments only when qualified by education or experience in the specific technical fields involved.
    b. Engineers shall not affix their signatures to any plans or documents dealing with subject matter in which they lack competence, nor to any plan or document not prepared under their direction and control.
    c. Engineers may accept assignments and assume responsibility for coordination of an entire project and sign and seal the engineering documents for the entire project, provided that each technical segment is signed and sealed only by the qualified engineers who prepared the segment.

3. Engineers shall issue public statements only in an objective and truthful manner.
   a. Engineers shall be objective and truthful in professional reports, statements, or testimony. They shall include all relevant and pertinent information in such reports, statements, or testimony, which should bear the date indicating when it was current.
   b. Engineers may express publicly technical opinions that are founded upon knowledge of the facts and competence in the subject matter.
   c. Engineers shall issue no statements, criticisms, or arguments on technical matters that are inspired or paid for by interested parties, unless they have prefaced their comments by explicitly identifying the interested parties on whose behalf they are speaking, and by revealing the existence of any interest the engineers may have in the matters.
4. Engineers shall act for each employer or client as faithful agents or trustees.
   a. Engineers shall disclose all known or potential conflicts of interest that could influence or appear to influence their judgment or the quality of their services.
   b. Engineers shall not accept compensation, financial or otherwise, from more than one party for services on the same project, or for services pertaining to the same project, unless the circumstances are fully disclosed and agreed to by all interested parties.
   c. Engineers shall not solicit or accept financial or other valuable consideration, directly or indirectly, from outside agents in connection with the work for which they are responsible.
   d. Engineers in public service as members, advisors, or employees of a governmental or quasi-governmental body or department shall not participate in decisions with respect to services solicited or provided by them or their organizations in private or public engineering practice.
   e. Engineers shall not solicit or accept a contract from a governmental body on which a principal or officer of their organization serves as a member.
5. Engineers shall avoid deceptive acts.
   a. Engineers shall not falsify their qualifications or permit misrepresentation of their or their associates' qualifications. They shall not misrepresent or exaggerate their responsibility in or for the subject matter of prior assignments. Brochures or other presentations incident to the solicitation of employment shall not misrepresent pertinent facts concerning employers, employees, associates, joint venturers, or past accomplishments.
   b. Engineers shall not offer, give, solicit, or receive, either directly or indirectly, any contribution to influence the award of a contract by public authority, or which may be reasonably construed by the public as having the effect or intent of influencing the awarding of a contract. They shall not offer any gift or other valuable consideration in order to secure work. They shall not pay a commission, percentage, or brokerage fee in order to secure work, except to a bona fide employee or bona fide established commercial or marketing agencies retained by them.

III. Professional Obligations

1. Engineers shall be guided in all their relations by the highest standards of honesty and integrity.
   a. Engineers shall acknowledge their errors and shall not distort or alter the facts.
   b. Engineers shall advise their clients or employers when they believe a project will not be successful.
   c. Engineers shall not accept outside employment to the detriment of their regular work or interest. Before accepting any outside engineering employment, they will notify their employers.

    d. Engineers shall not attempt to attract an engineer from another employer by false or misleading pretenses.

    e. Engineers shall not promote their own interest at the expense of the dignity and integrity of the profession.

2. Engineers shall at all times strive to serve the public interest.

    a. Engineers are encouraged to participate in civic affairs; career guidance for youths; and work for the advancement of the safety, health, and well-being of their community.

    b. Engineers shall not complete, sign, or seal plans and/or specifications that are not in conformity with applicable engineering standards. If the client or employer insists on such unprofessional conduct, they shall notify the proper authorities and withdraw from further service on the project.

    c. Engineers are encouraged to extend public knowledge and appreciation of engineering and its achievements.

    d. Engineers are encouraged to adhere to the principles of sustainable development[1] in order to protect the environment for future generations.

3. Engineers shall avoid all conduct or practice that deceives the public.

    a. Engineers shall avoid the use of statements containing a material misrepresentation of fact or omitting a material fact.

    b. Consistent with the foregoing, engineers may advertise for recruitment of personnel.

    c. Consistent with the foregoing, engineers may prepare articles for the lay or technical press, but such articles shall not imply credit to the author for work performed by others.

4. Engineers shall not disclose, without consent, confidential information concerning the business affairs or technical processes of any present or former client or employer, or public body on which they serve.

    a. Engineers shall not, without the consent of all interested parties, promote or arrange for new employment or practice in connection with a specific project for which the engineer has gained particular and specialized knowledge.

    b. Engineers shall not, without the consent of all interested parties, participate in or represent an adversary interest in connection with a specific project or proceeding in which the engineer has gained particular specialized knowledge on behalf of a former client or employer.

5. Engineers shall not be influenced in their professional duties by conflicting interests.

    a. Engineers shall not accept financial or other considerations, including free engineering designs, from material or equipment suppliers for specifying their product.

    b. Engineers shall not accept commissions or allowances, directly or indirectly, from contractors or other parties dealing with clients or employers of the engineer in connection with work for which the engineer is responsible.

6. Engineers shall not attempt to obtain employment or advancement or professional engagements by untruthfully criticizing other engineers, or by other improper or questionable methods.

    a. Engineers shall not request, propose, or accept a commission on a contingent basis under circumstances in which their judgment may be compromised.

    b. Engineers in salaried positions shall accept part-time engineering work only to the extent consistent with policies of the employer and in accordance with ethical considerations.

    c. Engineers shall not, without consent, use equipment, supplies, laboratory, or office facilities of an employer to carry on outside private practice.

7. Engineers shall not attempt to injure, maliciously or falsely, directly or indirectly, the professional reputation, prospects, practice, or employment of other engineers. Engineers who believe others are guilty of unethical or illegal practice shall present such information to the proper authority for action.

   a. Engineers in private practice shall not review the work of another engineer for the same client, except with the knowledge of such engineer, or unless the connection of such engineer with the work has been terminated.

   b. Engineers in governmental, industrial, or educational employ are entitled to review and evaluate the work of other engineers when so required by their employment duties.

   c. Engineers in sales or industrial employ are entitled to make engineering comparisons of represented products with products of other suppliers.

8. Engineers shall accept personal responsibility for their professional activities, provided, however, that engineers may seek indemnification for services arising out of their practice for other than gross negligence, where the engineer's interests cannot otherwise be protected.

   a. Engineers shall conform with state registration laws in the practice of engineering.

   b. Engineers shall not use association with a nonengineer, a corporation, or partnership as a "cloak" for unethical acts.

9. Engineers shall give credit for engineering work to those to whom credit is due, and will recognize the proprietary interests of others.

   a. Engineers shall, whenever possible, name the person or persons who may be individually responsible for designs, inventions, writings, or other accomplishments.

   b. Engineers using designs supplied by a client recognize that the designs remain the property of the client and may not be duplicated by the engineer for others without express permission.

   c. Engineers, before undertaking work for others in connection with which the engineer may make improvements, plans, designs, inventions, or other records that may justify copyrights or patents, should enter into a positive agreement regarding ownership.

   d. Engineers' designs, data, records, and notes referring exclusively to an employer's work are the employer's property. The employer should indemnify the engineer for use of the information for any purpose other than the original purpose.

   e. Engineers shall continue their professional development throughout their careers and should keep current in their specialty fields by engaging in professional practice, participating in continuing education courses, reading in the technical literature, and attending professional meetings and seminars.

Footnote 1 "Sustainable development" is the challenge of meeting human needs for natural resources, industrial products, energy, food, transportation, shelter, and effective waste management while conserving and protecting environmental quality and the natural resource base essential for future development.

As Revised July 2007

By order of the United States District Court for the District of Columbia, former Section 11(c) of the NSPE Code of Ethics prohibiting competitive bidding, and all policy statements, opinions, rulings or other guidelines interpreting its scope, have been rescinded as unlawfully interfering with the legal right of engineers, protected under the antitrust laws, to provide price information to prospective clients; accordingly, nothing contained in the NSPE Code of Ethics, policy statements, opinions, rulings or other guidelines prohibits the submission of price quotations or competitive bids for engineering services at any time or in any amount.

Statement by NSPE Executive Committee

In order to correct misunderstandings which have been indicated in some instances since the issuance of the Supreme Court decision and the entry of the Final Judgment, it is noted that in its decision of April 25, 1978, the Supreme Court of the United States declared: "The Sherman Act does not require competitive bidding."

It is further noted that as made clear in the Supreme Court decision:

1. Engineers and firms may individually refuse to bid for engineering services.
2. Clients are not required to seek bids for engineering services.
3. Federal, state, and local laws governing procedures to procure engineering services are not affected, and remain in full force and effect.
4. State societies and local chapters are free to actively and aggressively seek legislation for professional selection and negotiation procedures by public agencies.
5. State registration board rules of professional conduct, including rules prohibiting competitive bidding for engineering services, are not affected and remain in full force and effect. State registration boards with authority to adopt rules of professional conduct may adopt rules governing procedures to obtain engineering services.
6. As noted by the Supreme Court, "nothing in the judgment prevents NSPE and its members from attempting to influence governmental action . . ."

NOTE: In regard to the question of application of the Code to corporations vis-à-vis real persons, business form or type should not negate nor influence conformance of individuals to the Code. The Code deals with professional services, which services must be performed by real persons. Real persons in turn establish and implement policies within business structures. The Code is clearly written to apply to the Engineer, and it is incumbent on members of NSPE to endeavor to live up to its provisions. This applies to all pertinent sections of the Code.

Part 4: Inventor's Notebook

INTRODUCTION

Using a Laboratory Notebook to record ideas, inventions, experimentation records, observations, and all work details is a vital part of any laboratory process. Careful attention to how you keep your Laboratory Notebook can have a positive impact on the patent outcome of a pending discovery or invention.

Following are some overall recommendations to help you keep more efficient and accurate Laboratory Notebook entries. Remember, however, that these are simply a suggested set of guidelines. Only your attorney can supply the exact guidelines he or she would like you to follow to satisfy specific legal requirements. That is why we recommend that you consult your legal counsel.

RECORDING DATA

Your Laboratory Notebook is a vital record of your work whether it is for patent purposes, legal records, or documenting drug research under FDA guidelines. The Laboratory Notebook can help you prove:

a. Exact details and dates of conception
b. Details and dates of reduction to practice
c. Diligence in reducing your invention to practice
d. Details regarding the structure and operation of your invention
e. Experimentation observations and results
f. A chronological record of your work
g. Other work details

Follow a few simple rules of thumb

1. Always record entries legibly, neatly, and in permanent ink.
2. Immediately enter into your notebook and date all original concepts, data, and observations, using separate headings to differentiate each.
3. Record all concepts, results, references, and other information in a systematic and orderly manner. (Language, charts, and numbering systems should be maintained consistently throughout.)
4. It is acceptable to make your entries brief. Always, however, include enough details for someone else to successfully duplicate the work you have recorded.
5. Label all figures and calculations.
6. Never, under any circumstances, remove pages from your Notebook.

Remember to treat your Laboratory Notebook as a legal document: It records the chronological history of your activities. The following guidelines should help you maintain the consistent and accurate entries needed for future legal purposes.

1. Start entries at the top of the first page, and always make successive, dated entries, working your way to the bottom of the last page.
2. After completing a page, sign it before continuing to the next page.
3. Make sure that you record the date of each entry clearly and unambiguously.
4. Never let anyone other than yourself write in your Notebook (excluding witness signatures, discussed later).
5. Never leave blank spaces, and never erase or remove material you have added. Simply draw lines through any blank spaces at the same time you are making your entries.

6. Do not erase errors. Just draw a single line through any erroneous entry, and then add your initials. Enter the correct entry nearby.
7. You can supplement your entries with supporting material (e.g., test-result printouts and other documentation). But you must permanently affix the material onto a page in its proper chronological location.
8. Never rely solely on any supplemental attachment. Always include your own entry describing the attachment and add any conclusions that you might draw from its substance.
9. Occasionally, secondary sources might be too large or inappropriate to attach directly to your Notebook. In this case, you can add all secondary sources to an ancillary record maintained precisely for this purpose. However, always remember to write a description of these secondary sources, clearly and unambiguously, in your Notebook.

Documenting Patent Activities

A primary purpose of a Laboratory Notebook is the support of documenting work that may be patentable. To support patent activities, it is necessary to provide clear, concise, chronological entries with specific dates. To rely on these dates, you must have at least one non-inventor corroborate that the events actually happened and that he or she understood your invention by signing and dating the "Disclosed to and Understood by" signature blocks.

Your Laboratory Notebook should help you document and prove:

1. *Conception Date*--The date that you knew your invention would solve the problem.
2. *Date of reduction to practice*--The moment that you made a working embodiment of your invention.
3. *Diligence in reducing your invention to practice*--Diligence refers to your intent and conscious effort to make a working embodiment. You are not required to rush, or even to take the most efficient development strategy. But your Notebook must include details relating to your diligent activities. These are dates and facts that show what activities you have conducted to reduce the invention to practice, and when such activities were conducted. Since you may still be diligent despite periods of not working on reducing your invention to practice, always remember to provide reasonable excuses for these periods of inactivity by supplying facts relating to why there was no activity during the period in question. (e.g., unavailability of test conditions or equipment).
4. *How to make and use your invention*--provide documentation details sufficient to teach a colleague how to make and use your invention.
5. *The best mode of practicing your invention*--document the best way to practice your invention.

A non-inventor colleague should corroborate each of these events/facts by signing the "Disclosed to and Understood by" on the relevant pages (as applicable).

Page _____ of 30
Date _____
Invention / Project Title _____
Entry Made By (Print Name) _____
Signature of Person Making Entry _____

Disclosed to and Understood by _____
Continued on Page _____

Page _____ of 30
Date _____
Invention / Project Title _____
Entry Made By (Print Name) _____
Signature of Person Making Entry _____

Disclosed to and Understood by _____
Continued on Page _____

Page _____ of 30
Date _____
Invention / Project Title _____
Entry Made By (Print Name) _____
Signature of Person Making Entry _____

Disclosed to and Understood by _____
Continued on Page _____

Page _____ of 30
Date _____
Invention / Project Title _____
Entry Made By (Print Name) _____
Signature of Person Making Entry _____

Disclosed to and Understood by _____
Continued on Page _____

Page _____ of 30
Date _____
Invention / Project Title _____
Entry Made By (Print Name) _____
Signature of Person Making Entry _____

Disclosed to and Understood by _____
Continued on Page _____

Page _____ of 30
Date _____
Invention / Project Title _____
Entry Made By (Print Name) _____
Signature of Person Making Entry _____

Disclosed to and Understood by _____
Continued on Page _____

Page _____ of 30
Date _____
Invention / Project Title _____
Entry Made By (Print Name) _____
Signature of Person Making Entry _____

Disclosed to and Understood by _____
Continued on Page _____

Page _____ of 30
Date _____
Invention / Project Title _____
Entry Made By (Print Name) _____
Signature of Person Making Entry _____

Disclosed to and Understood by _____
Continued on Page _____

Page _____ of 30
Date _____
Invention / Project Title _____
Entry Made By (Print Name) _____
Signature of Person Making Entry _____

Disclosed to and Understood by _____
Continued on Page _____

Page _____ of 30
Date _____
Invention / Project Title _____
Entry Made By (Print Name) _____
Signature of Person Making Entry _____

Disclosed to and Understood by _____
Continued on Page _____

Page _____ of 30
Date _____
Invention / Project Title _____
Entry Made By (Print Name) _____
Signature of Person Making Entry _____

Disclosed to and Understood by _____
Continued on Page _____

Page _____ of 30
Date _____
Invention / Project Title _____
Entry Made By (Print Name) _____
Signature of Person Making Entry _____

Disclosed to and Understood by _____
Continued on Page _____

Page _____ of 30
Date _____
Invention / Project Title _____
Entry Made By (Print Name) _____
Signature of Person Making Entry _____

Disclosed to and Understood by _____
Continued on Page _____

Page _____ of 30
Date _____
Invention / Project Title _____
Entry Made By (Print Name) _____
Signature of Person Making Entry _____

Disclosed to and Understood by _____

Continued on Page _____

Page _____ of 30
Date _____
Invention / Project Title _____
Entry Made By (Print Name) _____
Signature of Person Making Entry _____

Disclosed to and Understood by _____
Continued on Page _____

Page _____ of 30
Date _____
Invention / Project Title _____
Entry Made By (Print Name) _____
Signature of Person Making Entry _____

Disclosed to and Understood by _____

Continued on Page _____

Page _____ of 30
Date _____
Invention / Project Title _____
Entry Made By (Print Name) _____
Signature of Person Making Entry _____

Disclosed to and Understood by _____
Continued on Page _____

Page _____ of 30
Date _____
Invention / Project Title _____
Entry Made By (Print Name) _____
Signature of Person Making Entry _____

Disclosed to and Understood by _____

Continued on Page _____

Page _____ of 30
Date _____
Invention / Project Title _____
Entry Made By (Print Name) _____
Signature of Person Making Entry _____

Disclosed to and Understood by _____
Continued on Page _____

Page _____ of 30
Date _____
Invention / Project Title _____
Entry Made By (Print Name) _____
Signature of Person Making Entry _____

Disclosed to and Understood by _____
Continued on Page _____

Page _____ of 30
Date _____
Invention / Project Title _____
Entry Made By (Print Name) _____
Signature of Person Making Entry _____

Disclosed to and Understood by _____
Continued on Page _____

Page _____ of 30
Date _____
Invention / Project Title _____
Entry Made By (Print Name) _____
Signature of Person Making Entry _____

Disclosed to and Understood by _____
Continued on Page _____

Page _____ of 30
Date _____
Invention / Project Title _____
Entry Made By (Print Name) _____
Signature of Person Making Entry _____

Disclosed to and Understood by _____
Continued on Page _____

Page _____ of 30
Date _____
Invention / Project Title _____
Entry Made By (Print Name) _____
Signature of Person Making Entry _____

Disclosed to and Understood by _____
Continued on Page _____

Page _____ of 30
Date _____
Invention / Project Title _____
Entry Made By (Print Name) _____
Signature of Person Making Entry _____

Disclosed to and Understood by _____
Continued on Page _____

Page _____ of 30
Date _____
Invention / Project Title _____
Entry Made By (Print Name) _____
Signature of Person Making Entry _____

Disclosed to and Understood by _____
Continued on Page _____

Page _____ of 30
Date _____
Invention / Project Title _____
Entry Made By (Print Name) _____
Signature of Person Making Entry _____

Disclosed to and Understood by _____
Continued on Page _____

Page _____ of 30
Date _____
Invention / Project Title _____
Entry Made By (Print Name) _____
Signature of Person Making Entry _____

Disclosed to and Understood by _____
Continued on Page _____

Page _____ of 30
Date _____
Invention / Project Title _____
Entry Made By (Print Name) _____
Signature of Person Making Entry _____

Disclosed to and Understood by _____
Continued on Page _____

Page _____ of 30
Date _____
Invention / Project Title _____
Entry Made By (Print Name) _____
Signature of Person Making Entry _____

Disclosed to and Understood by _____